**Act fast,
be smart,
and have fun**

A
HANDBOOK
OF
CALCULATING
24

Victor Zhang

What's included in this book?

This book acts as an ultimate index of calculating 24, using 4 numbers each ranging from 1 to 13 inclusive. There are 1,362 such questions in total that have at least one solution of calculating 24.

How to use this book?

Each page is split into 2 columns. All questions are highlighted in bold text, followed by solution(s). In each question, 4 numbers are automatically listed in the ascending order.

If you want to find the solution to 2, 4, 3, and 8, you should firstly sort them into ascending order as 2, 3, 4, and 8, and then find 2 3 4 8 in this book. The solution to this question is listed right after:

$(8-4)×3×2$ or $(4+2-3)×8$ or $(8/2+4)×3$

Rules of calculating 24

Please follow the following rules to solve each question:

- Each of the 4 numbers should be used
- Each of the 4 numbers should be only used once
- Only the following 4 operations could be used: Addition, Subtraction, Multiplication, and Division
- Parenthesis could be used to change the calculating order if needed.

So for the numbers of 1, 2, 4, and 6, you could do:
$4 \times 6 \times (2 - 1) = 24$ or $(6 + 2) \times (4 - 1) = 24$

1 1 1 8

$(1+1+1)\times8$

1 1 1 11

$(11+1)\times(1+1)$

1 1 1 12

$(1+1)\times12\times1$

1 1 1 13

$(13-1)\times(1+1)$

1 1 2 6

$(1+1)\times6\times2$

$(2+1+1)\times6$

1 1 2 7

$(7+1)\times(2+1)$

1 1 2 8

$(2+1)\times8\times1$

1 1 2 9

$(9-1)\times(2+1)$

1 1 2 10

$(10+2)\times(1+1)$

$(10+1+1)\times2$

1 1 2 11

$11\times2+1+1$

$(1+1)\times11+2$

$(11+1)\times2\times1$

1 1 2 12

$12\times2+1-1$

$12/(1-1/2)$

1 1 2 13

$13\times2-1-1$

$(1+1)\times13-2$

$(13-1)\times2\times1$

1 1 3 4

$(1+1)\times4\times3$

1 1 3 5

$(5+1)\times(3+1)$

1 1 3 6

(6+1+1)×3

(3+1)×6×1

1137

(7-1)×(3+1)

(7+1)×3×1

1138

8×3+1-1

1139

(9+3)×(1+1)

(9-1)×3×1

1 1 3 10

(10-1-1)×3

1 1 3 11

(11+1)×(3-1)

1 1 3 12

(3-1)×12×1

1 1 3 13

(13-1)×(3-1)

1 1 4 4

(4+1+1)×4

1 1 4 5

(4+1)×5-1

(5+1)×4×1

1 1 4 6

6×4+1-1

1 1 4 7

(7-1)×4×1

(7+1)×(4-1)

1 1 4 8

(8+4)×(1+1)

(8-1-1)×4

(4-1)×8×1

1 1 4 9

(9-1)×(4-1)

1 1 4 10

(1+1)×10+4

1 1 4 12

   12×4/(1+1)

   (4-1-1)×12

1 1 5 5

   (5+1)×(5-1)

   (5×5-1)×1

1 1 5 6

   (5-1)×6×1

   (6-1)×5-1

1 1 5 7

   (7+5)×(1+1)

   (7-1)×(5-1)

1 1 5 8

   (5-1-1)×8

1 1 6 6

   (6+6)×(1+1)

   (6-1-1)×6

1 1 6 8

   8×6/(1+1)

1 1 6 9

   (1+1)×9+6

1 1 6 12

   (1+1)×6+12

1 1 7 10

   (1+1)×7+10

1 1 8 8

   (1+1)×8+8

1 1 9 13

   13+9+1+1

1 1 10 12

   12+10+1+1

1 1 10 13

   (13+10+1)×1

1 1 11 11

   11+11+1+1

1 1 11 12

(12+11+1)×1

1 1 11 13

13+11+1-1

1 1 12 12

12+12+1-1

1 1 12 13

(13+12-1)×1

1 1 13 13

13+13-1-1

1 2 2 4

(2+1)×4×2

1 2 2 5

(5+1)×(2+2)

(5+1)×2×2

1 2 2 6

(6+2)×(2+1)

6×2×2×1

(2+2)×6×1

1 2 2 7

(7-1)×(2+2)

(7-1)×2×2

1 2 2 8

(2+2-1)×8

(2×2-1)×8

1 2 2 9

(9+2+1)×2

1 2 2 10

(10-2)×(2+1)

(10+2)×2×1

(10+1)×2+2

1 2 2 11

(11+2-1)×2

(11×2+2)×1

1 2 2 12

(2-1)×12×2

(12+1)×2-2

(12-1)×2+2

(2/2+1)×12

1 2 2 13

(13+1-2)×2

(13×2-2)×1

1 2 3 3

(3+1)×3×2

1 2 3 4

(3+2+1)×4

4×3×2×1

(4+2)×(3+1)

1 2 3 5

(5+3)×(2+1)

(5+2+1)×3

(5-1)×3×2

(3+2)×5-1

1 2 3 6

(3+2-1)×6

(6+2)×3×1

(3-1)×6×2

1 2 3 7

7×3+2+1

(2+1)×7+3

(7+2-1)×3

1 2 3 8

(2-1)×8×3

(8-2)×(3+1)

(8+3+1)×2

8/(1-2/3)

1 2 3 9

9×3-2-1

(2+1)×9-3

(9+3)×2×1

(9+1-2)×3

1 2 3 10

10×2+3+1

(10-2)×3×1

(10+3-1)×2

(10+2)×(3-1)

1 2 3 11

(11-3)×(2+1)

(11-2-1)×3

11×2+3-1

(3-1)×11+2

1 2 3 12

(3+1-2)×12

(3+1)×12/2

12/(3/2-1)

1 2 3 13

13×2+1-3

(3-1)×13-2

1 2 4 4

(4+4)×(2+1)

(4+2)×4×1

(4-1)×4×2

1 2 4 5

(5+2-1)×4

(4+2)×(5-1)

1 2 4 6

(2-1)×6×4

(6+2)×(4-1)

1 2 4 7

(7+1-2)×4

(7+4+1)×2

1 2 4 8

(8+4)×2×1

(4+1-2)×8

(8-2)×4×1

(4/2+1)×8

1 2 4 9

(9-2-1)×4

(9+4-1)×2

(9+1)×2+4

1 2 4 10

(10-2)×(4-1)

(10×2+4)×1

(10/2+1)×4

1 2 4 11

(11+1)×(4-2)

(11+1)×4/2

(11-1)×2+4

1 2 4 12

(12-4)×(2+1)

(2+1)×4+12

12×4×1/2

(4-2)×12×1

12/(1-2/4)

1 2 4 13

(13+1)×2-4

(13-1)×(4-2)

(13-1)×4/2

1 2 5 5

5×5+1-2

1 2 5 6

(5+1-2)×6

(5+1)×(6-2)

(6+5+1)×2

1 2 5 7

(7+5)×2×1

(7+1)×(5-2)

(7-2)×5-1

1 2 5 8

(5+1)×8/2

(8+5-1)×2

(8-2)×(5-1)

(5-2)×8×1

1 2 5 9

(2+1)×5+9

9×2+5+1

(9-1)×(5-2)

1 2 5 10

10×2+5-1

10×5/2-1

1 2 5 12

(5-2-1)×12

(5+1)×2+12

(5-1)×12/2

1 2 5 13

(13-5)×(2+1)

5×2+13+1

1 2 6 6

(2+1)×6+6

(6+6)×2×1

(6-2)×6×1

(6/2+1)×6

1 2 6 7

(7-2-1)×6

(7+6-1)×2

(7+1)×6/2

(7-1)×(6-2)

1 2 6 8

(6-2-1)×8

8×6×1/2

(8+1)×2+6

1 2 6 9

(9-1)×6/2

(9×2+6)×1

1 2 6 10

(2+1)×10-6

(6+1)×2+10

(10-1)×2+6

(10/2-1)×6

1 2 6 11

6×2+11+1

1 2 6 12

12×6/(2+1)

(6×2+12)×1

(6/2-1)×12

1 2 6 13

6×2+13-1

1 2 7 7

(7×7-1)/2

1 2 7 8

8×2+7+1

(7+1)×2+8

(7-1)×8/2

1 2 7 9

9×2+7-1

7×2+9+1

1 2 7 10

(7×2+10)×1

1 2 7 11

7×2+11-1

1 2 7 12

(7-1)×2+12

1 2 8 8

(8×2+8)×1

(8/2-1)×8

1 2 8 9

9×8/(2+1)

8×2+9-1

(9-1)×2+8

1 2 8 10

(8-1)×2+10

1 2 8 13

13+8+2+1

1 2 9 11

(2+1)×11-9

1 2 9 12

12+9+2+1

9

1 2 9 13

    (13+9+2)×1

1 2 10 11

    11+10+2+1

1 2 10 12

    (12+10+2)×1

1 2 10 13

    13+10+2-1

1 2 11 11

    (11+11+2)×1

1 2 11 12

    12+11+2-1

    2/(1-11/12)

1 2 11 13

    (13+11)×(2-1)

1 2 12 12

    (2+1)×12-12

    (12+12)×(2-1)

1 2 12 13

    13+12+1-2

    2/(13/12-1)

1 2 13 13

    (13+13-2)×1

1 3 3 3

    (3+3)×(3+1)

    (3×3-1)×3

1 3 3 4

    (4+3+1)×3

    (3-1)×4×3

    (3+3)×4×1

1 3 3 5

    (5+3)×3×1

    (3+3)×(5-1)

1 3 3 6

    (6+3-1)×3

    (6+1)×3+3

1 3 3 7

(7×3+3)×1

1 3 3 8

(8+1)×3-3

(8-1)×3+3

1 3 3 9

(9-3)×(3+1)

(9+3)×(3-1)

(3-1/3)×9

(9×3-3)×1

1 3 3 10

(10+1-3)×3

(10-1)×3-3

1 3 3 11

(11-3)×3×1

1 3 3 12

(3+1)×3+12

(12-3-1)×3

(3/3+1)×12

1 3 4 4

(4+4)×3×1

(4+3-1)×4

1 3 4 5

5×4+3+1

(3+1)×5+4

(5+4-1)×3

(5+3)×(4-1)

1 3 4 6

6/(1-3/4)

1 3 4 7

7×4-3-1

(3+1)×7-4

7×3+4-1

(4-1)×7+3

1 3 4 8

(3+1)×4+8

(8+1-3)×4

(8+4)×(3-1)

8/(4/3-1)

1 3 4 9

(4+1)×3+9

(9-3)×4×1

9×3+1-4

(4-1)×9-3

1 3 4 10

(10-4)×(3+1)

(10-3-1)×4

(3-1)×10+4

1 3 4 11

(11+1-4)×3

(11-3)×(4-1)

4×3+11+1

1 3 4 12

(12-4)×3×1

(4+1-3)×12

12×4/(3-1)

(4×3+12)×1

1 3 4 13

(13-4-1)×3

4×3+13-1

1 3 5 6

6×3+5+1

(5+1)×3+6

1 3 5 7

(7+5)×(3-1)

(5+1)×(7-3)

1 3 5 8

(5+1-3)×8

5×3+8+1

(8-3)×5-1

1 3 5 9

(9-3)×(5-1)

(5×3+9)×1

(5/3+1)×9

1 3 5 10

10×3-5-1

5×3+10-1

1 3 5 11

(11-5)×(3+1)

(11+1)×(5-3)

1 3 5 12

(5+1)×12/3

(12+1-5)×3

(5-1)×3+12

(5-3)×12×1

1 3 5 13

(13-5)×3×1

(13-1)×(5-3)

1 3 6 6

(6+1-3)×6

(6+6)×(3-1)

(6×3+6)×1

1 3 6 7

(7-3)×6×1

(7+1)×(6-3)

6×3+7-1

(7-1)×3+6

1 3 6 8

(8-3-1)×6

8×6/(3-1)

(6-3)×8×1

(6/3+1)×8

1 3 6 9

(3-1)×9+6

(6-1)×3+9

(9+1)×3-6

(9-1)×(6-3)

(9/3+1)×6

1 3 6 10

    (10×3-6)×1

1 3 6 11

    (11+1)×6/3

    (11-1)×3-6

1 3 6 12

    (12-6)×(3+1)

    (6-3-1)×12

    12×6×1/3

    (3-1)×6+12

    12/(1-3/6)

1 3 6 13

    (13+1-6)×3

    (13-1)×6/3

1 3 7 7

    (7-1)×(7-3)

1 3 7 8

    (7-3-1)×8

3/(1-7/8)

1 3 7 9

    (7+1)×9/3

1 3 7 10

    (3-1)×7+10

    10×3+1-7

1 3 7 12

    (7-1)×12/3

1 3 7 13

    13+7+3+1

    (13-7)×(3+1)

1 3 8 8

    (3+1)×8-8

    (3-1)×8+8

    (8+1)×8/3

1 3 8 9

    9×8×1/3

    3/(9/8-1)

1 3 8 10

$(10-1)\times8/3$

1 3 8 11

$11\times3-8-1$

1 3 8 12

$12+8+3+1$

$12\times8/(3+1)$

$(12/3-1)\times8$

1 3 8 13

$(13+8+3)\times1$

1 3 9 9

$(9-1)\times9/3$

1 3 9 10

$(10+1)\times3-9$

1 3 9 11

$11+9+3+1$

$(11\times3-9)\times1$

$(11/3-1)\times9$

1 3 9 12

$(3+1)\times9-12$

$(12-1)\times3-9$

$(12+9+3)\times1$

$(9/3-1)\times12$

1 3 9 13

$13+9+3-1$

1 3 10 10

$10+10+3+1$

1 3 10 11

$11\times3+1-10$

$(11+10+3)\times1$

1 3 10 12

$12+10+3-1$

1 3 11 11

$11+11+3-1$

1 3 11 12

$12\times3-11-1$

(11+1)×3-12

1 3 12 12

(12×3-12)×1

1 3 12 13

12×3+1-13

(13-1)×3-12

1 3 13 13

13+13+1-3

1 4 4 4

(4+1)×4+4

(4+4)×(4-1)

1 4 4 5

(5×4+4)×1

1 4 4 6

(6+1)×4-4

(6-1)×4+4

1 4 4 7

4×4+7+1

(7×4-4)×1

1 4 4 8

(8-1)×4-4

(4×4+8)×1

1 4 4 9

(9+1-4)×4

4×4+9-1

1 4 4 10

(10-4)×4×1

1 4 4 11

(11-4-1)×4

1 4 4 12

(12-4)×(4-1)

(4-1)×4+12

(4/4+1)×12

1 4 5 5

5×4+5-1

(5-1)×5+4

1 4 5 6

6/(5/4-1)

4/(1-5/6)

1 4 5 7

7×4+1-5

(5-1)×7-4

1 4 5 8

(5+1)×(8-4)

(5-1)×4+8

1 4 5 9

(4-1)×5+9

(9-4)×5-1

1 4 5 10

(10+1-5)×4

(10-4)×(5-1)

1 4 5 11

(11-5)×4×1

1 4 5 12

(5+1-4)×12

(12-5-1)×4

1 4 5 13

(13-5)×(4-1)

1 4 6 6

(4+1)×6-6

(4-1)×6+6

1 4 6 7

(7+1-4)×6

4/(7/6-1)

1 4 6 8

(6+1-4)×8

(8-4)×6×1

8/(1-4/6)

1 4 6 9

(9-4-1)×6

1 4 6 10

(4-1)×10-6

1 4 6 11

(11+1-6)×4

(11+1)×(6-4)

1 4 6 12

(12-6)×4×1

12×6/(4-1)

(6-4)×12×1

12/(6/4-1)

(12/4+1)×6

1 4 6 13

13+6+4+1

(13-6-1)×4

(13-1)×(6-4)

1 4 7 7

(7+1)×(7-4)

1 4 7 8

8×4-7-1

(7+1)×4-8

(7-1)×(8-4)

(7-4)×8×1

1 4 7 9

(9-1)×(7-4)

1 4 7 11

(4+1)×7-11

1 4 7 12

12+7+4+1

(7-4-1)×12

(7+1)×12/4

(12+1-7)×4

1 4 7 13

(13-7)×4×1

(13+7+4)×1

1 4 8 8

(8-4-1)×8

(8×4-8)×1

(8/4+1)×8

1 4 8 9

> 9×8/(4-1)

> 8×4+1-9

> (9-1)×4-8

1 4 8 11

> 11+8+4+1

> (11+1)×8/4

1 4 8 12

> 12×8×1/4

> (8+1)×4-12

> (12+8+4)×1

> 12/(1-4/8)

1 4 8 13

> 13+8+4-1

> (13+1-8)×4

> (13-1)×8/4

1 4 9 10

> 10+9+4+1

1 4 9 11

> (4-1)×11-9

> 9×4-11-1

> (11+9+4)×1

1 4 9 12

> 12+9+4-1

> (9-1)×12/4

> (9×4-12)×1

1 4 9 13

> 9×4+1-13

1 4 10 10

> (10+10+4)×1

> 10×10/4-1

1 4 10 11

> 11+10+4-1

1 4 10 12

> 12×10/(4+1)

> (10-1)×4-12

4/(1-10/12)

1 4 12 12

(4-1)×12-12

(12/4-1)×12

1 5 5 5

(5-1/5)×5

1 5 5 6

6×5-5-1

(5+1)×5-6

1 5 5 9

(5+1)×(9-5)

1 5 5 10

(10-5)×5-1

1 5 5 11

(11-5)×(5-1)

1 5 5 12

(5/5+1)×12

1 5 5 13

13+5+5+1

1 5 6 6

(6×5-6)×1

1 5 6 7

6×5+1-7

(7-1)×5-6

1 5 6 8

(8+1-5)×6

1 5 6 9

(9-5)×6×1

1 5 6 10

(5+1)×(10-6)

(10-5-1)×6

1 5 6 11

(6+1)×5-11

(11-6)×5-1

1 5 6 12

12+6+5+1

(5+1)×6-12

(6+1-5)×12

(12-6)×(5-1)

1 5 6 13

(13+6+5)×1

1 5 7 8

(7+1-5)×8

(7+1)×(8-5)

1 5 7 9

(7-1)×(9-5)

1 5 7 10

7×5-10-1

(7/5+1)×10

1 5 7 11

11+7+5+1

(5+1)×(11-7)

(11+1)×(7-5)

(7×5-11)×1

1 5 7 12

(7-5)×12×1

7×5+1-12

(12+7+5)×1

(12-7)×5-1

1 5 7 13

13+7+5-1

(13-7)×(5-1)

(13-1)×(7-5)

1 5 8 8

(5-1)×8-8

(8-5)×8×1

1 5 8 9

(9-5-1)×8

(9-1)×(8-5)

9/(1-5/8)

1 5 8 10

10+8+5+1

(10/5+1)×8

1 5 8 11

(8-1)×5-11

(11+8+5)×1

1 5 8 12

(5+1)×(12-8)

(8-5-1)×12

12+8+5-1

12×8/(5-1)

1 5 8 13

(13-8)×5-1

1 5 9 9

9+9+5+1

1 5 9 10

(10+9+5)×1

1 5 9 11

11+9+5-1

1 5 9 12

(5-1)×9-12

(9+1)×12/5

1 5 9 13

(5+1)×(13-9)

1 5 10 10

10+10+5-1

1 5 10 11

(11+1)×10/5

1 5 10 12

12×10×1/5

12/(1-5/10)

1 5 10 13

(13-1)×10/5

1 5 11 11

(11×11-1)/5

1 5 11 12

(11-1)×12/5

1 5 12 12

12×12/(5+1)

1 6 6 6

(6-1)×6-6

1 6 6 8

6/(1-6/8)

1 6 6 9

(9+1-6)×6

1 6 6 10

(10-6)×6×1

1 6 6 11

11+6+6+1

(11-6-1)×6

6×6-11-1

1 6 6 12

(12+6+6)×1

(6×6-12)×1

(6/6+1)×12

1 6 6 13

13+6+6-1

6×6+1-13

1 6 7 9

(7+1)×(9-6)

1 6 7 10

10+7+6+1

(10+1-7)×6

(7-1)×(10-6)

1 6 7 11

(11-7)×6×1

(6-1)×7-11

(11+7+6)×1

1 6 7 12

12+7+6-1

(7+1-6)×12

(12-7-1)×6

(7-1)×6-12

1 6 8 8

(8+1-6)×8

8/(8/6-1)

1 6 8 9

9+8+6+1

(9-6)×8×1

8/(1-6/9)

1 6 8 10

(10-6-1)×8

(10+8+6)×1

6/(10/8-1)

1 6 8 11

11+8+6-1

(11+1-8)×6

(11+1)×(8-6)

1 6 8 12

(12-8)×6×1

(8-6)×12×1

(12/6+1)×8

1 6 8 13

(13-8-1)×6

(13-1)×(8-6)

1 6 9 9

(9-1)×(9-6)

(9+9+6)×1

1 6 9 10

10+9+6-1

(10/6+1)×9

1 6 9 12

(9-6-1)×12

(12+1-9)×6

12/(9/6-1)

6/(1-9/12)

1 6 9 13

(13-9)×6×1

1 6 10 12

12×10/(6-1)

1 6 10 13

(13+1-10)×6

1 6 11 12

(11+1)×12/6

1 6 11 13

(13×11+1)/6

1 6 12 12

12×12×1/6

12/(1-6/12)

1 6 12 13

(13-1)×12/6

1 7 7 9

9+7+7+1

1 7 7 10

(7+1)×(10-7)

(10+7+7)×1

1 7 7 11

11+7+7-1

(7-1)×(11-7)

1 7 7 12

(7/7+1)×12

1 7 8 8

8+8+7+1

1 7 8 9

(9+1-7)×8

(9+8+7)×1

1 7 8 10

10+8+7-1

(10-7)×8×1

1 7 8 11

(7+1)×(11-8)

(11-7-1)×8

1 7 8 12

(8+1-7)×12

(7-1)×(12-8)

1 7 9 9

9+9+7-1

1 7 9 10

(9-1)×(10-7)

1 7 9 11

(11+1)×(9-7)

1 7 9 12

(7+1)×(12-9)

(9-7)×12×1

1 7 9 13

(7-1)×(13-9)

(13-1)×(9-7)

1 7 10 12

(10-7-1)×12

10/(1-7/12)

1 7 10 13

(7+1)×(13-10)

1 7 12 12

12×12/(7-1)

1 7 12 13

(13+1)×12/7

1 7 13 13

(13×13-1)/7

1 8 8 8

(8+8+8)×1

1 8 8 9

9+8+8-1

1 8 8 10

(10+1-8)×8

1 8 8 11

(11-8)×8×1

1 8 8 12

(12-8-1)×8

(8/8+1)×12

8/(1-8/12)

1 8 9 11

(11+1-9)×8

(9-1)×(11-8)

9/(11/8-1)

1 8 9 12

(12-9)×8×1

(9+1-8)×12

8/(12/9-1)

1 8 9 13

(13-9-1)×8

1 8 10 11

(11+1)×(10-8)

1 8 10 12

(12+1-10)×8

(10-8)×12×1

1 8 10 13

(13-10)×8×1

(13-1)×(10-8)

1 8 11 12

(11-8-1)×12

1 8 11 13

(13+1-11)×8

1 8 12 12

12/(12/8-1)

1 9 9 12

(9-1)×(12-9)

(9/9+1)×12

1 9 10 12

(10+1-9)×12

1 9 10 13

(9-1)×(13-10)

1 9 11 11

(11+1)×(11-9)

1 9 11 12

(11-9)×12×1

1 9 11 13

(13-1)×(11-9)

1 9 12 12

(12-9-1)×12

1 10 10 12

(10/10+1)×12

1 10 11 12

(11+1-10)×12

(11+1)×(12-10)

1 10 12 12

(12-10)×12×1

1 10 12 13

(13-10-1)×12

(13-1)×(12-10)

1 11 11 12

(11/11+1)×12

1 11 11 13

(11+1)×(13-11)

(13/11+1)×11

1 11 12 12

(12+1-11)×12

1 11 12 13

(13-11)×12×1

1 11 13 13

(13-1)×(13-11)

(11/13+1)×13

1 12 12 12

(12/12+1)×12

1 12 12 13

(13+1-12)×12

1 12 13 13

(13/13+1)×12

2 2 2 3

(2+2)×3×2

3×2×2×2

2 2 2 4

(4+2)×(2+2)

(2+2+2)×4

(4+2)×2×2

(2×2+2)×4

2 2 2 5

(5×2+2)×2

2 2 2 7

(7×2-2)×2

2 2 2 8

(8-2)×(2+2)

(8+2+2)×2

(8-2)×2×2

(2×2+8)×2

(2/2+2)×8

2 2 2 9

(9+2)×2+2

2 2 2 10

10×2+2+2

10×2+2×2

2 2 2 11

(2/2+11)×2

(11+2)×2-2

2 2 2 12

(2+2)×12/2

(2×2-2)×12

12×2+2-2

2 2 2 13

(13-2/2)×2

(13-2)×2+2

2 2 3 3

(3+3)×(2+2)

(3+3)×2×2

(3×2+2)×3

2 2 3 4

(4+2+2)×3

(2×2+4)×3

2 2 3 5

(5×2-2)×3

2 2 3 6

(2/2+3)×6

(6-2)×3×2

(3×2-2)×6

(3×2+6)×2

2 2 3 7

(2/2+7)×3

(7+3+2)×2

2 2 3 8

8×3+2-2

(8+3)×2+2

2 2 3 9

(9-3)×(2+2)

(9-3)×2×2

(9-2/2)×3

9×2+3×2

(2/3+2)×9

2 2 3 10

(10+3)×2-2

2 2 3 11

(11+3-2)×2

2 2 3 12

(2+2)×3+12

(12-2-2)×3

3×2×2+12

(12-2×2)×3

(3-2/2)×12

(3-2)×12×2

12/(2-3/2)

(12/2+2)×3

2 2 3 13

(13+2-3)×2

2 2 4 4

(4×2-2)×4

(4×2+4)×2

2 2 4 5

5×4+2+2

(2+2)×5+4

5×4+2×2

5×2×2+4

(2/2+5)×4

(5-2)×4×2

2 2 4 6

6×4+2-2

(4+2)×(6-2)

(6+4+2)×2

(4-2)×6×2

(4/2+2)×6

2 2 4 7

7×4-2-2

(2+2)×7-4

7×4-2×2

7×2×2-4

(7-2/2)×4

(7+4)×2+2

2 2 4 8

(2+2)×4+8

4×2×2+8

(4-2/2)×8

(4+2)×8/2

8×2+4×2

(8+2)×2+4

(8×2-4)×2

(8/2+2)×4

2 2 4 9

9×2+4+2

(9+4)×2-2

2 2 4 10

(10-4)×(2+2)

(10-2-2)×4

(10-4)×2×2

(10-2×2)×4

(10+4-2)×2

(10+2)×(4-2)

(10+2)×4/2

(4/2+10)×2

2 2 4 11

11×2+4-2

(4-2)×11+2

11×4/2+2

11×2+4/2

2 2 4 12

(4+2)×2+12

(12+2)×2-4

(12-2)×2+4

2 2 4 13

13×2+2-4

(4-2)×13-2

13×4/2-2

13×2-4/2

2 2 5 5

5×5-2/2

(5+5+2)×2

2 2 5 6

(5-2/2)×6

(6+2)×(5-2)

(6+5)×2+2

2 2 5 7

7×2+5×2

2 2 5 8

(8+5)×2-2

2 2 5 9

(9+5-2)×2

2 2 5 10

(5+2)×2+10

(10-2)×(5-2)

(2/5+2)×10

(10×5-2)/2

2 2 5 11

(11-5)×(2+2)

(11-5)×2×2

2 2 5 12

5×2+12+2

12/(5/2-2)

2 2 6 6

(6+2)×6/2

6×2+6×2

2 2 6 7

(7+2)×2+6

(7+6)×2-2

2 2 6 8

(8-2-2)×6

(8-2×2)×6

8×2+6+2

(6+2)×2+8

(8+6-2)×2

(8-2)×(6-2)

2 2 6 9

(6/2+9)×2

(9×2-6)×2

2 2 6 10

6×2+10+2

10×2+6-2

(10-2)×6/2

2 2 6 11

(11-2)×2+6

2 2 6 12

(12-6)×(2+2)

(6-2-2)×12

(12-6)×2×2

(6-2×2)×12

(6-2)×12/2

(12/2-2)×6

(12/2+6)×2

2 2 6 13

(13+2)×2-6

2 2 7 7

(7+7-2)×2

2 2 7 8

(7-2-2)×8

(7-2×2)×8

7×2+8+2

2 2 7 10

(10/2+7)×2

2 2 7 12

7×2+12-2

2 2 7 13

13+7+2+2

(13-7)×(2+2)

2×2+13+7

(13-7)×2×2

2 2 8 8

(2+2)×8-8

8×2×2-8

(8-2)×8/2

(8/2+8)×2

2 2 8 9

9×2+8-2

2 2 8 10

8×2+10-2

10×2+8/2

(10×2-8)×2

(10-2)×2+8

(10/2-2)×8

2 2 8 12

12+8+2+2

12×8/(2+2)

12×8/2/2

2×2+12+8

(8-2)×2+12

(8/2-2)×12

2 2 9 10

   (9-2)×2+10

2 2 9 11

   11+9+2+2

   2×2+11+9

2 2 9 12

   (2+2)×9-12

   9×2×2-12

   9×2+12/2

2 2 10 10

   10+10+2+2

   2×2+10+10

2 2 10 11

   (11×2-10)×2

2 2 10 13

   2/2+13+10

2 2 11 11

   (2/11+2)×11

2 2 11 12

   2/2+12+11

2 2 11 13

   13+11+2-2

2 2 12 12

   12+12+2-2

   (12×2-12)×2

2 2 12 13

   13+12-2/2

2 2 13 13

   (2-2/13)×13

2 3 3 3

   (3+3+2)×3

   (3×3+3)×2

2 3 3 5

   (5+2)×3+3

   (5×3-3)×2

2 3 3 6

6×3+3×2

3×3×2+6

(3+3-2)×6

(3+3)×(6-2)

(6+3+3)×2

2 3 3 7

(7-3)×3×2

(7+3-2)×3

(7+2)×3-3

2 3 3 8

(3×2-3)×8

(3-2)×8×3

(3+3)×8/2

(3/3+2)×8

2 3 3 9

(3+2)×3+9

(9+2-3)×3

9×2+3+3

(9-2)×3+3

2 3 3 10

10×3-3×2

(10/2+3)×3

2 3 3 11

(11-2)×3-3

(3/3+11)×2

2 3 3 12

12×2+3-3

(3+3)×2+12

2 3 3 13

(13-3-2)×3

(13+3)×3/2

3×3+13+2

(13-3/3)×2

2 3 4 4

(3+2)×4+4

4×4×3/2

(4-2)×4×3

2 3 4 5

(5+3-2)×4

(5+4+3)×2

2 3 4 6

(3-2)×6×4

6×3+4+2

(4+2)×3+6

(6-3)×4×2

(6+4-2)×3

(4/2+6)×3

6×2+4×3

(6×2-4)×3

(6/2+3)×4

2 3 4 7

(7+2-3)×4

(4+2)×(7-3)

(7+3)×2+4

2 3 4 8

(8-4)×3×2

(4+2-3)×8

(8/2+4)×3

2 3 4 9

9×4×2/3

(9+3)×(4-2)

(9+3)×4/2

2 3 4 10

10×3-4-2

(10+2-4)×3

(10-4/2)×3

4×3+10+2

(4+3)×2+10

2 3 4 11

(11-3-2)×4

(11+4-3)×2

(11+3)×2-4

2 3 4 12

    (3×2-4)×12

    (12-3×2)×4

    (12+4)×3/2

    (4+2)×12/3

    (4-3)×12×2

    (12/3+2)×4

2 3 4 13

    4×2+13+3

    (13+3-4)×2

    (13-3)×2+4

2 3 5 5

    5×5+2-3

    (5+5-2)×3

    (5+3)×(5-2)

2 3 5 6

    6×5-3×2

    5×3×2-6

    (5+2-3)×6

    (5-3)×6×2

    (5+3)×6/2

    (6/2+5)×3

2 3 5 7

    7×3+5-2

    (5-2)×7+3

    5×3+7+2

2 3 5 8

    8×2+5+3

    (5+3)×2+8

    8/(2-5/3)

2 3 5 9

    (9-5)×3×2

    9×3+2-5

    (5-2)×9-3

    (9×5+3)/2

2 3 5 10

(10+2)×(5-3)

(10+5-3)×2

2 3 5 11

(11+5)×3/2

5×2+11+3

(11+2-5)×3

(11-3)×(5-2)

11×2+5-3

5×3+11-2

(5-3)×11+2

2 3 5 12

12/(3-5/2)

2 3 5 13

3×2+13+5

13×2+3-5

(5-3)×13-2

2 3 6 6

(3+2)×6-6

6×6×2/3

(6+2)×(6-3)

(6+3)×2+6

(6×3-6)×2

(6/3+2)×6

2 3 6 7

7×6/2+3

7×3+6/2

(7×2-6)×3

2 3 6 8

(8+2)×3-6

6×3+8-2

(8-2)×3+6

2 3 6 9

(9-3-2)×6

(6+2)×9/3

6×2+9+3

(9-3)×(6-2)

(3-2/6)×9	6×3×2-12
9×6/2-3	12×3/2+6
9×3-6/2	12×3-6×2
(9+6-3)×2	(12+2-6)×3
2 3 6 10	(6-2)×3+12
(10-6)×3×2	(12-2)×3-6
(10-3×2)×6	6×3+12/2
(10+6)×3/2	(12+3)×2-6
(10+2)×6/3	(12-3)×2+6
(10-2)×(6-3)	**2 3 6 13**
(6/3+10)×2	13+6+3+2
2 3 6 11	13×2-6/3
(11-3)×6/2	13×6/3-2
(11-6/2)×3	**2 3 7 7**
11×2+6/3	7×2+7+3
11×6/3+2	**2 3 7 8**
2 3 6 12	(7+2)×8/3
3×2+12+6	(8-2)×(7-3)

(8+7-3)×2

8/(7/3-2)

2 3 7 9

(9+7)×3/2

(7-2)×3+9

(7×3-9)×2

2 3 7 10

10×2+7-3

(10×7+2)/3

2 3 7 11

(3+2)×7-11

3×2+11+7

(11-7)×3×2

11×3-7-2

2 3 7 12

12+7+3+2

(7-3-2)×12

12/(7/2-3)

(7-3)×12/2

2 3 7 13

7×2+13-3

(13+2-7)×3

2 3 8 8

(8-3-2)×8

(8+8)×3/2

(8×2-8)×3

2 3 8 9

(9-3×2)×8

(9-3)×8/2

2 3 8 10

3×2+10+8

10×3+2-8

2 3 8 11

11+8+3+2

8×2+11-3

(11-2)×8/3

(11-3)×2+8

2 3 8 12

(12-8)×3×2

(8-3×2)×12

8×3/2+12

(8-2)×12/3

(12-8/2)×3

(12/2-3)×8

(8×3-12)×2

(12/3+8)×2

2 3 8 13

(13+3)×2-8

2 3 9 9

3×2+9+9

(9+2)×3-9

9×2+9-3

(9/3+9)×2

2 3 9 10

10+9+3+2

10×3/2+9

(9×2-10)×3

(10-2)×9/3

2 3 9 12

(9-3)×2+12

2 3 9 13

(13-9)×3×2

(13-2)×3-9

(13×3+9)/2

2 3 10 10

(10-3)×2+10

2 3 10 12

12×10/(3+2)

12×3-10-2

(10+2)×3-12

10×2+12/3

(10×2-12)×3

(10/2-3)×12

10×3-12/2

2 3 10 13

13+10+3-2

(13-10/2)×3

2 3 11 11

11×3+2-11

2 3 11 12

12+11+3-2

2 3 11 13

(13+11)×(3-2)

2 3 12 12

12×12/3/2

(12+12)×(3-2)

(12×3+12)/2

(12/3-2)×12

2 3 12 13

13+12+2-3

2 3 13 13

13×3-13-2

2 4 4 4

4×4+4×2

(4+4-2)×4

(4/2+4)×4

(4+4+4)×2

(4×4-4)×2

2 4 4 5

(5+2)×4-4

(5×2-4)×4

(4+4)×(5-2)

2 4 4 6

(4×2-4)×6

4×4+6+2

(4+4)×6/2

(6+4)×2+4

2 4 4 7

(7-4)×4×2

(7-2)×4+4

2 4 4 8

(4+2)×(8-4)

8×4-4×2

4×4×2-8

(8+2-4)×4

(8+4)×(4-2)

(8+4)×4/2

(8-4/2)×4

8×2+4+4

(4+4)×2+8

(4/4+2)×8

2 4 4 9

(9-2)×4-4

2 4 4 10

(4-2)×10+4

10×4/2+4

4×4+10-2

(10+4)×2-4

2 4 4 11

(4/4+11)×2

(11×4+4)/2

2 4 4 12

(12-4-2)×4

4×2+12+4

12×4/(4-2)

(4-4/2)×12

12×2+4-4

2 4 4 13

(13-4/4)×2

(13×4-4)/2

2 4 5 5

(5+5)×2+4

2 4 5 6

6×5-4-2

(4+2)×5-6

5×4+6-2

(6-2)×5+4

(5+4)×2+6

2 4 5 7

(7+5)×(4-2)

(7+5)×4/2

2 4 5 8

(8-5)×4×2

(4×2-5)×8

(5+2-4)×8

(5-4/2)×8

8×5/2+4

5×4+8/2

(5×4-8)×2

2 4 5 9

(4+2)×(9-5)

(9+2-5)×4

(9+5)×2-4

2 4 5 10

5×2+10+4

2 4 5 11

4×2+11+5

(11+5-4)×2

2 4 5 12

(12-4)×(5-2)

(5-2)×4+12

(5-4)×12×2

2 4 5 13

13+5+4+2

(13-5-2)×4

(13+4-5)×2

2 4 6 6

(6+2-4)×6

(6+6)×(4-2)

(6+6)×4/2

(6-4/2)×6

(6-4)×6×2

(6×2-6)×4

2 4 6 7

(6+2)×(7-4)

7×4+2-6

(6-2)×7-4

7×2+6+4

6/(2-7/4)

2 4 6 8

8×6/(4-2)

8×6×2/4

8×4-6-2

(6+2)×4-8

6×2+8+4

(6-2)×4+8

(8/4+2)×6

(8+6)×2-4

2 4 6 9

(9-6)×4×2

(4-2)×9+6

9×4/2+6

9×4-6×2

(9-6/2)×4

(4/6+2)×9

6/(9/4-2)

2 4 6 10

(4+2)×(10-6)

(10-4-2)×6

4×2+10+6

(10+2-6)×4

(10-4)×(6-2)

(10+2)×(6-4)

(10+6-4)×2

2 4 6 11

11×2+6-4

(6-4)×11+2

(11+4)×2-6

4/(2-11/6)

2 4 6 12

12+6+4+2

(4+2)×6-12

(4×2-6)×12

(12-4×2)×6

(4-2)×6+12

6×4/2+12

(6+2)×12/4

(12-4)×6/2

(6×4-12)×2

12/(2-6/4)

2 4 6 13

13×2+4-6

(6-4)×13-2

(13-4)×2+6

4/(13/6-2)

2 4 7 7

(7+7)×2-4

2 4 7 8

(7×2-8)×4

8×7/2-4

7×4-8/2

2 4 7 9

4×2+9+7

(9+7-4)×2

2 4 7 10

(10-7)×4×2

(4-2)×7+10

7×4/2+10

(10-2)×(7-4)

2 4 7 11

11+7+4+2

(4+2)×(11-7)

(11+2-7)×4

2 4 7 12

(7+2)×4-12

12/(4-7/2)

2 4 8 8

4×2+8+8

(4-2)×8+8

8×4/2+8

(8-2)×(8-4)

(8+8-4)×2

2 4 8 9

(9-4-2)×8

2 4 8 10

10+8+4+2

10×4-8×2

(8×2-10)×4

(10-4)×8/2

(10-8/2)×4

(10+2)×8/4

10×2+8-4

8×4+2-10

(10-2)×4-8

(8/4+10)×2

(10×4+8)/2

2 4 8 11

(11-8)×4×2

(11-4×2)×8

11×2+8/4

11×8/4+2

2 4 8 12

(4+2)×(12-8)

(8-4-2)×12

8×2+12-4

(12+2-8)×4

(8-4)×12/2

(12+4)×2-8

(12-4)×2+8

2 4 8 13

13×2-8/4

13×8/4-2

2 4 9 9

9+9+4+2

2 4 9 10

9×2+10-4

9×4-10-2

2 4 9 12

(12-9)×4×2

(9×2-12)×4

12/(9/2-4)

(9×4+12)/2

(12/4+9)×2

2 4 9 13

(4+2)×(13-9)

13+9+4-2

4/2+13+9

(13+2-9)×4

2 4 10 10

(4/10+2)×10

2 4 10 11

11×4-10×2

(11-10/2)×4

(11-4)×2+10

2 4 10 12

(10-4×2)×12

12+10+4-2

4/2+12+10

(10-2)×12/4

(10-4)×2+12

12/(10/4-2)

2 4 10 13

(13-10)×4×2

(13+4)×2-10

2 4 11 11

11+11+4-2

4/2+11+11

2 4 11 12

(11-2)×4-12

2 4 12 12

12×12/(4+2)

12×4-12×2

(12/2-4)×12

(12-12/2)×4

2 4 13 13

13+13+2-4

13+13-4/2

2 5 5 7

7×2+5+5

2 5 5 8

(5/5+2)×8

2 5 5 9

5×2+9+5

(5-2)×5+9

2 5 5 10

(5-2/10)×5

2 5 5 11

(5+2)×5-11

(5/5+11)×2

2 5 5 12

12+5+5+2

12×2+5-5

2 5 5 13

(13-5)×(5-2)

(5×5-13)×2

(13-5/5)×2

2 5 6 6

(5×2-6)×6

(5-2)×6+6

2 5 6 7

(7+2-5)×6

6×2+7+5

(7-5)×6×2

2 5 6 8

5×2+8+6

(6+2-5)×8

(6+2)×(8-5)

6×5+2-8

(8-2)×5-6

2 5 6 9

6×5/2+9

2 5 6 10

(5-2)×10-6

10×6×2/5

(10+5)×2-6

(10/5+2)×6

2 5 6 11

11+6+5+2

(11-5-2)×6

(11-5)×(6-2)

(11+6-5)×2

2 5 6 12

12×6/(5-2)

12×5/2-6

(5-6/2)×12

(6-5)×12×2

6×5-12/2

2 5 6 13

(13-5)×6/2

(13+5-6)×2

2 5 7 7

5×2+7+7

2 5 7 8

(5×2-7)×8

2 5 7 9

7×5-9-2

2 5 7 10

    10+7+5+2

    (10+2)×(7-5)

    (10+7-5)×2

2 5 7 11

    11×2+7-5

    (7-5)×11+2

    (11×5-7)/2

2 5 7 13

    13×2+5-7

    7×5+2-13

    (7×5+13)/2

    (7-5)×13-2

2 5 8 8

    8×5-8×2

    (8×5+8)/2

2 5 8 9

    9+8+5+2

    9×8/(5-2)

    (8-2)×(9-5)

    (9+8-5)×2

2 5 8 10

    (10-5-2)×8

    (10-2)×(8-5)

2 5 8 11

    (11-5)×8/2

    (11+5)×2-8

2 5 8 12

    (5×2-8)×12

    (8+2)×12/5

2 5 8 13

    (13-5×2)×8

    13+8+5-2

    8×2+13-5

    (13+2)×8/5

    (13-5)×2+8

2 5 9 10

    10×2+9-5

2 5 9 11

    (5-2)×11-9

    9×2+11-5

    (9-2)×5-11

2 5 9 12

    (9-5-2)×12

    12+9+5-2

    12/(5-9/2)

    (9-5)×12/2

2 5 10 10

    (10+2)×10/5

    (10/5+10)×2

2 5 10 11

    11+10+5-2

    11×2+10/5

    11×10/5+2

2 5 10 12

    (12+5)×2-10

    (12-5)×2+10

2 5 10 13

    10×5-13×2

    13×2-10/5

    13×10/5-2

2 5 11 12

    12/(11/2-5)

    (11-5)×2+12

2 5 12 12

    (12-5×2)×12

    (5-2)×12-12

    (12-2)×12/5

    (12×5-12)/2

2 5 12 13

    12/2+13+5

    (13+5)×2-12

2 6 6 6

6×2+6+6

6×6-6×2

6×6/2+6

2 6 6 7

(7-6/2)×6

(7-2)×6-6

(7×6+6)/2

2 6 6 8

(8-6)×6×2

(6×2-8)×6

(8+2-6)×6

(6-6/2)×8

(6/6+2)×8

2 6 6 9

(6+2)×(9-6)

(9+6)×2-6

(9×6-6)/2

2 6 6 10

10+6+6+2

10×6/2-6

6×6-10-2

2 6 6 11

(6/6+11)×2

2 6 6 12

(12-6-2)×6

(12-6)×(6-2)

12×2+6-6

(6×6+12)/2

(12/6+2)×6

2 6 6 13

(13-6/6)×2

2 6 7 8

(7+2-6)×8

(8+7)×2-6

2 6 7 9

9+7+6+2

(9-7)×6×2

(9+2-7)×6

7×6-9×2

2 6 7 10

(6+2)×(10-7)

(7×2-10)×6

2 6 7 11

(11+7-6)×2

2 6 7 12

(7-6)×12×2

2 6 7 13

13+7+6-2

(13-7)×(6-2)

(13-7-2)×6

(13+6-7)×2

2 6 8 8

8+8+6+2

(6-2)×8-8

(8-8/2)×6

2 6 8 9

(6×2-9)×8

9×8×2/6

2 6 8 10

(10-8)×6×2

(10+2-8)×6

(8-2)×(10-6)

(10+2)×(8-6)

(10+8-6)×2

(10+6)×2-8

8/(2-10/6)

2 6 8 11

(6+2)×(11-8)

(11-6-2)×8

11×2+8-6

(8-6)×11+2

2 6 8 12

12+8+6-2

12×8/(6-2)

(8×2-12)×6

(8-2)×6-12

(12-6)×8/2

(6-8/2)×12

8×6-12×2

2 6 8 13

6/2+13+8

13×2+6-8

(8-6)×13-2

2 6 9 9

(9+9-6)×2

(6/9+2)×9

2 6 9 10

(10-2)×(9-6)

(9-10/2)×6

2 6 9 11

(11-9)×6×2

11+9+6-2

11×6/2-9

(11+2-9)×6

2 6 9 12

(6+2)×(12-9)

(6-2)×9-12

6/2+12+9

9×2+12-6

12/(2-9/6)

2 6 10 10

10+10+6-2

10×2+10-6

2 6 10 11

6/2+11+10

(11+6)×2-10

2 6 10 12

(10-6-2)×12

(12-10)×6×2

(6×2-10)×12

(10+2)×12/6

(12+2-10)×6

(10-6)×12/2

(10-12/2)×6

(10×6-12)/2

(12/6+10)×2

2 6 10 13

(6+2)×(13-10)

10/2+13+6

(13-6)×2+10

2 6 11 12

11×2+12/6

12/(6-11/2)

12×11/6+2

2 6 11 13

(13-11)×6×2

(13+2-11)×6

2 6 12 12

12×6/2-12

12/2+12+6

(12+6)×2-12

(12-6)×2+12

2 6 12 13

13×2-12/6

12/(13/2-6)

13×12/6-2

2 7 7 8

8+7+7+2

(7/7+2)×8

2 7 7 10

(10/7+2)×7

2 7 7 11

(7-2)×7-11

(7/7+11)×2

2 7 7 12

12+7+7-2

12×2+7-7

2 7 7 13

(13-7/7)×2

2 7 8 8

(8+2-7)×8

(7-8/2)×8

(8×7-8)/2

2 7 8 9

(9+7)×2-8

2 7 8 11

(7×2-11)×8

11+8+7-2

(8-2)×(11-7)

(11+8-7)×2

2 7 8 12

(12-7-2)×8

(8-7)×12×2

2 7 8 13

(13-7)×8/2

8/2+13+7

(13+7-8)×2

2 7 9 10

10+9+7-2

(10+2)×(9-7)

(10+9-7)×2

2 7 9 11

11×2+9-7

(9-7)×11+2

2 7 9 13

9×2+13-7

13×2+7-9

(9-7)×13-2

2 7 10 10

(10-2)×(10-7)

(10+7)×2-10

2 7 10 11

10×7/2-11

10×2+11-7

2 7 10 12

12×10/(7-2)

10/2+12+7

(7-10/2)×12

2 7 11 12

(11-7-2)×12

(11-7)×12/2

12/2+11+7

(11+7)×2-12

2 7 12 12

(7×2-12)×12

(12+2)×12/7

2 7 12 13

12/(7-13/2)

(13-7)×2+12

2 8 8 8

8×8/2-8

(8+8)×2-8

(8/8+2)×8

2 8 8 9

(9+2-8)×8

2 8 8 10

10+8+8-2

(8-10/2)×8

2 8 8 11

(8/8+11)×2

2 8 8 12

(8-2)×(12-8)

8/2+12+8

12×2+8-8

2 8 8 13

(13-8-2)×8

(8×2-13)×8

(13-8/8)×2

2 8 9 9

9+9+8-2

(9/9+2)×8

2 8 9 10

(10+2-9)×8

(9+8)×2-10

2 8 9 11

8/2+11+9

(11+9-8)×2

2 8 9 12

9×8/2-12

(9-8)×12×2

(9-12/2)×8

(8/12+2)×9

2 8 9 13

(8-2)×(13-9)

(13+8-9)×2

9/(2-13/8)

2 8 10 10

8/2+10+10

(10+2)×(10-8)

(10+10-8)×2

(10/10+2)×8

2 8 10 11

(11+2-10)×8

(10-2)×(11-8)

10/2+11+8

11×2+10-8

(10-8)×11+2

2 8 10 12

10×2+12-8

12/2+10+8

(10+8)×2-12

2 8 10 13

13×2+8-10

(10-8)×13-2

2 8 11 11

(11/11+2)×8

2 8 11 12

(12+2-11)×8

2 8 12 12

(12-8-2)×12

12×12/(8-2)

(12-8)×12/2

(8-12/2)×12

12/(2-12/8)

(12/12+2)×8

2 8 12 13

(13+2-12)×8

2 8 13 13

(13/13+2)×8

2 9 9 11

(9/9+11)×2

2 9 9 12

12×2+9-9

12/2+9+9

(9+9)×2-12

2 9 9 13

(13-9/9)×2

2 9 10 10

10/2+10+9

2 9 10 11

(10+2)×(11-9)

(11+10-9)×2

2 9 10 12

(10-2)×(12-9)

(10-9)×12×2

2 9 10 13

10×2+13-9

(13+9-10)×2

2 9 11 11

11×2+11-9

(11-9)×11+2

2 9 11 13

13×2+9-11

(11-9)×13-2

2 9 12 13

(13-9-2)×12

(13-9)×12/2

2 9 13 13

(13+9)/2+13

2 10 10 11

(10/10+11)×2

2 10 10 12

(10+2)×(12-10)

12×2+10-10

2 10 10 13

(10-2)×(13-10)

(13-10/10)×2

2 10 11 11

(11+11-10)×2

2 10 11 12

11×2+12-10

(11-10)×12×2

(12-10)×11+2

2 10 11 13

(10+2)×(13-11)

(13+10-11)×2

2 10 12 13

13×2+10-12

(12+10)/2+13

(12-10)×13-2

2 11 11 11

(11/11+11)×2

2 11 11 12

12×2+11-11

2 11 11 13

11×2+13-11

(11+11)/2+13

(13-11/11)×2

(13-11)×11+2

2 11 12 12

(12-11)×12×2

(12/12+11)×2

2 11 12 13

(13+11-12)×2

(13+11)/2+12

2 11 13 13

13×2+11-13

(13-11)×13-2

(13+13)/2+11

(13/13+11)×2

2 12 12 12

12×2+12-12

(12+12)/2+12

2 12 12 13

(13-12)×12×2

(13-12/12)×2

2 12 13 13

12×2+13-13

2 13 13 13

(13-13/13)×2

3 3 3 3

3×3×3-3

3 3 3 4

(3×3-3)×4

(4+3)×3+3

3 3 3 5

5×3+3×3

3 3 3 6

6×3+3+3

(3+3)×3+6

(3/3+3)×6

(6+3)×3-3

3 3 3 7

(3+3)×(7-3)

(3/3+7)×3

3 3 3 8

8×3+3-3

3 3 3 9

(9-3/3)×3

3 3 3 10

10×3-3-3

(10-3)×3+3

3 3 3 11

11×3-3×3

3 3 3 12

(3+3)×12/3

3×3+12+3

(3-3/3)×12

(12-3)×3-3

3 3 4 4

4×3+4×3

(4×3-4)×3

3 3 4 5

(3/3+5)×4

(5-3)×4×3

(5+4)×3-3

3 3 4 6

6×4+3-3

3 3 4 7

(7-3/3)×4

(7+4-3)×3

3 3 4 8

(3+3)×(8-4)

(4-3/3)×8

(4-3)×8×3

3 3 4 9

    4×3+9+3

    (9+3-4)×3

    (9/3+3)×4

3 3 4 11

    3×3+11+4

    (11-4)×3+3

3 3 4 12

    (3+3-4)×12

    (12-3-3)×4

    4×3×3-12

    12×3-4×3

    (12/3+4)×3

3 3 4 13

    (13-4)×3-3

3 3 5 5

    5×5-3/3

3 3 5 6

6×5-3-3

    (3+3)×5-6

    (3×3-5)×6

    (5-3/3)×6

    (5+3)×(6-3)

    5×3+6+3

    (6+5-3)×3

3 3 5 7

    (5×3-7)×3

3 3 5 9

    (3+3)×(9-5)

    (5+3)×9/3

    (9+3)×(5-3)

    (9/3+5)×3

3 3 5 10

    3×3+10+5

    (10+3-5)×3

    (3-3/5)×10

3 3 5 12

5×3+12-3

(12-5)×3+3

3 3 5 13

13+5+3+3

13×3-5×3

3 3 6 6

(6/3+6)×3

3 3 6 7

7×3+6-3

(6-3)×7+3

(7+3)×3-6

3 3 6 8

(3×3-6)×8

(6+3)×8/3

3 3 6 9

3×3+9+6

6×3+9-3

9×3+3-6

(6-3)×9-3

(9+3)×6/3

(9-3)×3+6

3 3 6 10

(3+3)×(10-6)

(10-3-3)×6

(6×3-10)×3

(10-6/3)×3

3 3 6 11

11×3-6-3

(11+3-6)×3

(11-3)×(6-3)

3 3 6 12

12+6+3+3

(3+3)×6-12

3 3 6 13

(13-3×3)×6

(13-3)×3-6

(13-6)×3+3

3 3 7 7

(3/7+3)×7

3 3 7 8

3×3+8+7

3 3 7 9

7×3+9/3

(9-3)×(7-3)

9×7/3+3

3 3 7 11

11+7+3+3

(3+3)×(11-7)

3 3 7 12

(3×3-7)×12

(12+3-7)×3

(7-3)×3+12

3 3 7 13

(7×3-13)×3

3 3 8 8

8/(3-8/3)

3 3 8 9

(9-3-3)×8

(8+3)×3-9

(8-3)×3+9

3 3 8 10

10+8+3+3

8/(10/3-3)

3 3 8 12

(3+3)×(12-8)

(8-3-3)×12

(12-3×3)×8

(12-3)×8/3

3 3 8 13

(13+3-8)×3

3 3 9 9

9+9+3+3

9×3-9/3

(3-3/9)×9

9×9/3-3

3 3 9 10

10×3+3-9

3 3 9 11

(11-3)×9/3

(11-9/3)×3

3 3 9 12

12×3-9-3

(9+3)×3-12

(9-3)×12/3

3 3 9 13

(3+3)×(13-9)

3 3 10 13

3/3+13+10

3 3 11 12

(11-3×3)×12

3/3+12+11

11×3+3-12

3 3 11 13

13+11+3-3

3 3 12 12

12×12/(3+3)

12+12+3-3

(12-12/3)×3

3 3 12 13

13+12-3/3

13×3-12-3

3 4 4 4

(4+3)×4-4

3 4 4 5

(5+4-3)×4

4×4+5+3

3 4 4 6

(6-4)×4×3

(4×3-6)×4

(4-3)×6×4

(4+4)×(6-3)

(6/3+4)×4

(4/4+3)×6

3 4 4 7

(7+3-4)×4

(4/4+7)×3

3 4 4 8

4×3+8+4

8×3+4-4

(8-3)×4+4

(4×4-8)×3

3 4 4 9

9×4-4×3

(4-4/3)×9

(4+4)×9/3

(9-4/4)×3

3 4 4 10

(10-3)×4-4

3 4 4 11

4×4+11-3

3 4 4 12

(3-4/4)×12

(12/4+3)×4

3 4 4 13

13+4+4+3

(13-4-3)×4

3 4 5 5

5×5+3-4

5×3+5+4

3 4 5 6

(5+3-4)×6

3 4 5 7

4×3+7+5

(7-5)×4×3

(5+3)×(7-4)

5×4+7-3

(7-3)×5+4

(7+5-4)×3

3 4 5 8

8×4-5-3

(5+3)×4-8

(8+3-5)×4

(8+4)×(5-3)

(5-4)×8×3

(5+4)×8/3

3 4 5 9

(5×3-9)×4

(9+4-5)×3

3 4 5 10

10×4×3/5

(5-3)×10+4

3 4 5 11

(4+3)×5-11

11×3-5-4

3 4 5 12

12+5+4+3

(4+3-5)×12

(5+3)×12/4

12×4/(5-3)

12×5/3+4

5×4+12/3

(5×4-12)×3

(12/4+5)×3

3 4 5 13

(13+5)×4/3

5×3+13-4

3 4 6 6

4×3+6+6

6×6-4×3

(6+4)×3-6

(6+6-4)×3

3 4 6 8

(8-6)×4×3

(4×3-8)×6

(8+4)×6/3

(8-6/3)×4

(8/4+6)×3

3 4 6 9

(9+3-6)×4

(9+3)×(6-4)

3 4 6 10

6×3+10-4

10×6/3+4

(10+4-6)×3

(10-4)×3+6

3 4 6 11

11+6+4+3

(11-4-3)×6

6/(3-11/4)

3 4 6 12

12×4×3/6

(12+6)×4/3

(6+3)×4-12

(6×3-12)×4

(12-4)×(6-3)

(6-3)×4+12

(4-6/3)×12

3 4 6 13

(13+3)×6/4

6/(13/4-3)

3 4 7 7

7×3+7-4

7×4+3-7

(7-3)×7-4

(7-4)×7+3

3 4 7 8

(7-3)×4+8

3 4 7 9

(9-7)×4×3

9×3+4-7

(7+4)×3-9

(7-4)×9-3

3 4 7 10

10+7+4+3

(10+3-7)×4

(10-4)×(7-3)

3 4 7 11

(11+7)×4/3

(11-3)×(7-4)

(11+4-7)×3

3 4 7 12

7×3+12/4

12×7/3-4

7×4-12/3

12×7/4+3

3 4 8 9

9+8+4+3

(4×3-9)×8

(9+3)×8/4

(9-3)×(8-4)

3 4 8 10

(10-4-3)×8

(10-8)×4×3

(10+8)×4/3

(10-8/4)×3

3 4 8 11

(11+3-8)×4

8×4+3-11

(11-3)×4-8

8/(4-11/3)

3 4 8 12

12×4/3+8

12×4-8×3

12×3-8-4

(8+4)×3-12

(12+4-8)×3

(8-4)×3+12

3 4 8 13

(13-4)×8/3

8/(13/3-4)

3 4 9 9

(9+9)×4/3

9×4-9-3

(9-9/3)×4

(9-4)×3+9

3 4 9 11

(11-9)×4×3

(11×9-3)/4

3 4 9 12

(9-4-3)×12

9×4/3+12

9×3-12/4

(12+3-9)×4

(12-4)×9/3

12×9/4-3

(3-4/12)×9

3 4 9 13

(13+4-9)×3

3 4 10 10

10×3+4-10

3 4 10 12

(12-10)×4×3

(4×3-10)×12

(10-4)×12/3

(10-12/3)×4

12/(3-10/4)

3 4 10 13

13+10+4-3

(13+3-10)×4

10×4-13-3

3 4 11 12

12+11+4-3

(11-3)×12/4

(11-12/4)×3

3 4 11 13

(13-11)×4×3

(13+11)×(4-3)

11×3+4-13

13×3-11-4

3 4 12 12

(12+12)×(4-3)

(12-3)×4-12

3 4 12 13

13+12+3-4

3 5 5 6

(5+5)×3-6

(5/5+3)×6

3 5 5 7

(7+5)×(5-3)

(5/5+7)×3

3 5 5 8

(5+3)×(8-5)

8×3+5-5

3 5 5 9

(9-5/5)×3

(9/5+3)×5

3 5 5 11

11+5+5+3

3 5 5 12

(3-5/5)×12

3 5 6 6

(6+3-5)×6

(6+6)×(5-3)

3 5 6 7

(7+5)×6/3

(7+6-5)×3

3 5 6 8

8×6/(5-3)

(5-6/3)×8

(6-5)×8×3

3 5 6 9

(5+3)×(9-6)

(5-3)×9+6

(6-3)×5+9

6×5+3-9

(9-3)×5-6

(6+5)×3-9

(9+5-6)×3

3 5 6 10

10+6+5+3

(10/5+6)×3

3 5 6 11

(5×3-11)×6

6×3+11-5

(11-5)×3+6

3 5 6 12

(5+3-6)×12

(12-5-3)×6

(5-3)×6+12

3 5 6 13

(13-5)×(6-3)

3 5 7 8

7×3+8-5

7×5-8-3

(8-5)×7+3

3 5 7 9

9+7+5+3

9×5-7×3

(5-7/3)×9

(9+3)×(7-5)

3 5 7 10

(5+3)×(10-7)

(5-3)×7+10

(10+5-7)×3

3 5 7 11

(11-5)×(7-3)

(11×7-5)/3

3 5 7 12

(7+3)×12/5

12×3-7-5

(7+5)×3-12

3 5 7 13

(13×5+7)/3

3 5 8 8

8+8+5+3

(5-3)×8+8

3 5 8 9

9×3+5-8

(8-5)×9-3

3 5 8 11

(5+3)×(11-8)

(11-5-3)×8

(11-3)×(8-5)

(11+5-8)×3

3 5 8 12

(5×3-12)×8

(12+3)×8/5

3 5 8 13

8×5-13-3

3 5 9 9

9×5/3+9

(9-3)×(9-5)

3 5 9 10

(9+3)×10/5

(10-5)×3+9

3 5 9 12

    (5+3)×(12-9)

    (5-9/3)×12

    (12+5-9)×3

    (9-5)×3+12

3 5 9 13

    13+9+5-3

    (13-5)×9/3

    (13×9+3)/5

3 5 10 10

    (10-10/5)×3

3 5 10 11

    10×3+5-11

    (10-3)×5-11

3 5 10 12

    (10-5-3)×12

    12+10+5-3

3 5 10 13

    (5+3)×(13-10)

    13×3-10-5

    (13+5-10)×3

3 5 11 11

    11+11+5-3

3 5 11 12

    (11-5)×12/3

3 5 12 12

    12×5-12×3

    (12×5+12)/3

3 5 12 13

    (5×3-13)×12

    (13-3)×12/5

3 5 13 13

    13+13+3-5

3 6 6 6

    (6-3)×6+6

    (6+6)×6/3

(6-6/3)×6

(6/6+3)×6

3 6 6 7

7×6-6×3

(7+3-6)×6

(6/6+7)×3

3 6 6 8

8×3+6-6

(8-3)×6-6

3 6 6 9

9+6+6+3

9×6/3+6

6×6-9-3

(9-6/6)×3

3 6 6 10

(6-3)×10-6

3 6 6 11

(11×6+6)/3

3 6 6 12

6×3+12-6

12×6/(6-3)

6×6/3+12

12×3-6-6

(6+6)×3-12

(3-6/6)×12

(12-6)×3+6

(12/6+6)×3

3 6 6 13

(13-6-3)×6

(13×6-6)/3

3 6 7 7

(7+7-6)×3

(7/7+3)×6

3 6 7 8

8+7+6+3

(8+3-7)×6

(7-6)×8×3

3 6 7 9

7×3+9-6

(7-9/3)×6

(9+6-7)×3

(9-6)×7+3

3 6 7 10

7×6/3+10

3 6 7 12

(6+3-7)×12

(12-6)×(7-3)

3 6 7 13

6×3+13-7

(13-7)×3+6

3 6 8 8

8×6/3+8

8×6-8×3

(8/8+3)×6

3 6 8 9

9×8/(6-3)

(9+3-8)×6

(9+3)×(8-6)

(6-9/3)×8

3 6 8 10

(10+6-8)×3

3 6 8 12

(12-6-3)×8

(8-12/3)×6

3 6 8 13

13+8+6-3

3 6 9 9

9×3+6-9

(9-6)×9-3

(9/9+3)×6

3 6 9 10

(10+3-9)×6

(9-3)×(10-6)	**3 6 10 10**
10×9/3-6	(3-6/10)×10
9×6-10×3	(10/10+3)×6
(6-10/3)×9	**3 6 10 11**
3 6 9 11	11+10+6-3
(6-3)×11-9	(11+3-10)×6
(11-3)×(9-6)	**3 6 10 12**
(11+6-9)×3	6/3+12+10
(11-6)×3+9	10×3+6-12
3 6 9 12	10×6-12×3
12×6×3/9	(10×6+12)/3
12+9+6-3	(12+6-10)×3
(9+3)×12/6	(10-6)×3+12
(9-3)×6-12	(10-12/6)×3
3 6 9 13	**3 6 11 11**
6/3+13+9	6/3+11+11
(13+3)×9/6	(11/11+3)×6
13×3-9-6	**3 6 11 12**

(11-6-3)×12

(12+3-11)×6

3 6 11 13

(13+6-11)×3

3 6 12 12

(6-3)×12-12

(12-6)×12/3

(6-12/3)×12

(12/12+3)×6

3 6 12 13

(13+3-12)×6

3 6 13 13

13+13-6/3

(13/13+3)×6

3 7 7 7

7+7+7+3

(7/7+7)×3

3 7 7 8

8×3+7-7

3 7 7 9

(9-7/7)×3

3 7 7 10

7×3+10-7

(10-7)×7+3

3 7 7 12

(3-7/7)×12

3 7 7 13

13+7+7-3

(13-7)×(7-3)

3 7 8 8

(7-3)×8-8

(8-7)×8×3

(8/8+7)×3

3 7 8 9

(9+7-8)×3

3 7 8 11

81

7×3+11-8

(8-3)×7-11

(11-8)×7+3

3 7 8 12

(7+3-8)×12

12+8+7-3

12×8/(7-3)

(7-12/3)×8

3 7 8 13

(13-7-3)×8

13×3-8-7

3 7 9 9

(9+3)×(9-7)

(9×7+9)/3

(9/9+7)×3

3 7 9 10

9×3+7-10

(10+7-9)×3

(10-7)×9-3

3 7 9 11

11+9+7-3

(9-3)×(11-7)

3 7 9 12

7×3+12-9

(7-3)×9-12

(12-7)×3+9

(12-9)×7+3

3 7 9 13

9×7-13×3

(7-13/3)×9

3 7 10 10

10+10+7-3

(10/10+7)×3

3 7 10 11

(11-3)×(10-7)

(11+7-10)×3

3 7 10 13

7×3+13-10

10×3+7-13

(13-10)×7+3

3 7 11 11

(11/11+7)×3

3 7 11 12

(11+3)×12/7

(12+7-11)×3

(11-7)×3+12

3 7 12 12

(12-7-3)×12

(12×7-12)/3

(12/12+7)×3

3 7 12 13

(13-7)×12/3

12/3+13+7

(13+7-12)×3

3 7 13 13

(13/13+7)×3

3 8 8 8

8×3+8-8

(8×8+8)/3

3 8 8 9

(9-8)×8×3

(9-8/8)×3

3 8 8 10

(10×8-8)/3

3 8 8 11

11+8+8-3

3 8 8 12

12×8/3-8

(3-8/8)×12

3 8 9 9

8×3+9-9

3 8 9 10

(10-9)×8×3

10+9+8-3

(9+3)×(10-8)

3 8 9 11

9×3+8-11

(11-8)×9-3

3 8 9 12

(8+3-9)×12

(9-3)×(12-8)

3 8 9 13

9/3+13+8

(13-8)×3+9

3 8 10 10

8×3+10-10

3 8 10 11

(11-10)×8×3

3 8 10 12

12×10/(8-3)

3 8 11 11

8×3+11-11

(11-3)×(11-8)

3 8 11 12

(12-11)×8×3

3 8 12 12

8×3+12-12

12/3+12+8

(12-8)×3+12

3 8 12 13

(13-8-3)×12

(13-12)×8×3

(13+3)×12/8

3 8 13 13

8×3+13-13

3 9 9 9

9+9+9-3

(9×9-9)/3

(9-9/9)×3

3 9 9 10

(9+9-10)×3

3 9 9 11

(9+3)×(11-9)

11×9/3-9

3 9 9 12

9×3+9-12

9/3+12+9

(3-9/9)×12

(12-9)×9-3

3 9 9 13

(9-3)×(13-9)

3 9 10 10

(9-10/10)×3

3 9 10 11

9/3+11+10

(10+9-11)×3

3 9 10 12

(9+3)×(12-10)

(9+3-10)×12

3 9 10 13

9×3+10-13

(13-10)×9-3

3 9 11 11

(3-9/11)×11

(9-11/11)×3

3 9 11 12

(11-3)×(12-9)

12/3+11+9

(11+9-12)×3

3 9 11 13

(9+3)×(13-11)

3 9 12 12

12×12/(9-3)

12×9/3-12

(9-12/12)×3

3 9 12 13

(12+9-13)×3

(13-9)×3+12

3 9 13 13

(9-13/13)×3

3 10 10 12

12/3+10+10

(10+10-12)×3

(3-10/10)×12

3 10 11 12

(10+3-11)×12

3 10 11 13

(11-3)×(13-10)

(11+10-13)×3

3 11 11 12

(3-11/11)×12

3 11 12 12

(11+3-12)×12

3 12 12 12

(3-12/12)×12

3 12 12 13

(12+3-13)×12

3 12 13 13

(3-13/13)×12

4 4 4 4

4×4+4+4

4 4 4 5

(4/4+5)×4

4 4 4 6

6×4+4-4

4 4 4 7

(4+4)×(7-4)

(7-4/4)×4

4 4 4 8

8×4-4-4

(4+4)×4-8

(4-4/4)×8

(8/4+4)×4

4 4 4 9

(9-4)×4+4

4 4 4 10

10×4-4×4

(4×4-10)×4

4 4 4 11

(11-4)×4-4

4 4 4 12

12+4+4+4

(4+4)×12/4

4×4+12-4

4 4 5 5

5×5-4/4

(5+5-4)×4

(4/5+4)×5

4 4 5 6

(5-4/4)×6

(5-4)×6×4

4 4 5 7

(7+4-5)×4

4 4 5 8

(4+4)×(8-5)

(4+4-5)×8

8×5-4×4

5×4+8-4

(8-4)×5+4

4 4 5 10

(10-5)×4+4

(10/5+4)×4

4 4 5 11

11+5+4+4

11×4-5×4

4 4 5 12

(5+4)×4-12

(12-5)×4-4

4 4 5 13

4×4+13-5

4 4 6 8

(8+4-6)×4

(8+4)×(6-4)

4 4 6 9

(4+4)×(9-6)

9×4×4/6

4 4 6 10

10+6+4+4

(6-4)×10+4

4 4 6 11

(11-6)×4+4

4 4 6 12

(4+4-6)×12

(12-4-4)×6

(4×4-12)×6

12×4-6×4

12×4/(6-4)

(12+4)×6/4

(12/6+4)×4

4 4 6 13

(13-6)×4-4

4 4 7 7

(4-4/7)×7

4 4 7 8

7×4+4-8

(8-4)×7-4

4 4 7 9

9+7+4+4

(9+4-7)×4

4 4 7 10

(4+4)×(10-7)

4 4 7 12

(12-4)×(7-4)

(7-4)×4+12

(12-7)×4+4

4 4 7 13

13×4-7×4

4 4 8 8

8+8+4+4

(8+4)×8/4

(8-4)×4+8

(8-8/4)×4

4 4 8 9

9×4-8-4

4 4 8 10

(10+4-8)×4

(10-4)×(8-4)

10×8/4+4

4 4 8 11

(4+4)×(11-8)

(11-4-4)×8

4 4 8 12

12×4×4/8

8×4+4-12

(4-8/4)×12

(12-4)×4-8

4 4 8 13

(4×4-13)×8

(13-8)×4+4

4 4 9 11

(11+4-9)×4

4 4 9 12

(4+4)×(12-9)

(9-12/4)×4

4 4 10 10

(10×10-4)/4

4 4 10 12

(10-4-4)×12

10×4-12-4

(12+4-10)×4

4 4 10 13

(4+4)×(13-10)

4/4+13+10

4 4 11 12

4/4+12+11

4 4 11 13

13+11+4-4

(13+4-11)×4

4 4 12 12

12+12+4-4

(12-4)×12/4

4 4 12 13

13+12-4/4

(13-4)×4-12

4 5 5 5

5×5+4-5

(5/5+5)×4

4 5 5 6

6×4+5-5

4 5 5 7

(7-5/5)×4

4 5 5 8

(4-5/5)×8

4 5 5 9

5×4+9-5

(9-5)×5+4

4 5 5 10

10+5+5+4

4 5 6 6

(6-5)×6×4

(6/6+5)×4

4 5 6 7

(7+5)×(6-4)

(7+5-6)×4

4 5 6 8

    (5+4-6)×8

4 5 6 9

    9+6+5+4

4 5 6 10

    5×4+10-6

    6×5+4-10

    (10-4)×5-6

    (10-6)×5+4

4 5 6 11

    (11+5)×6/4

4 5 6 12

    (6+4)×12/5

4 5 6 13

    (13-5-4)×6

4 5 7 7

    7×5-7-4

    (7/7+5)×4

4 5 7 8

    8+7+5+4

    (8+4)×(7-5)

    (7+5)×8/4

    (8+5-7)×4

4 5 7 9

    7×4+5-9

    (7-4)×5+9

    9×4-7-5

    (9-5)×7-4

4 5 7 10

    (7-5)×10+4

4 5 7 11

    5×4+11-7

    (11-7)×5+4

4 5 7 12

    (5+4-7)×12

    12×4/(7-5)

4 5 7 13

(13-5)×(7-4)

(13×7+5)/4

4 5 8 8

(5-8/4)×8

(8/8+5)×4

4 5 8 9

(9+5-8)×4

(9-5)×4+8

4 5 8 10

(8+4)×10/5

(4-8/5)×10

(8-10/5)×4

(8/10+4)×5

4 5 8 11

(11-5)×(8-4)

(11+4)×8/5

4 5 8 12

(12-5-4)×8

5×4+12-8

8×5-12-4

(12-4)×(8-5)

(8-5)×4+12

(12-8)×5+4

4 5 8 13

8×4+5-13

(13-5)×4-8

4 5 9 9

(9/9+5)×4

4 5 9 10

(10-4)×(9-5)

(10+5-9)×4

4 5 9 12

12×5/4+9

12×5-9×4

4 5 9 13

5×4+13-9

(13-9)×5+4

4 5 10 10

10×10/5+4

(10/10+5)×4

4 5 10 11

10×4-11-5

(11+5-10)×4

4 5 10 12

12×5×4/10

(4-10/5)×12

4 5 10 13

13+10+5-4

4 5 11 11

(11-4)×5-11

(11/11+5)×4

4 5 11 12

(11-5-4)×12

12+11+5-4

(12+5-11)×4

4 5 11 13

(13+11)×(5-4)

4 5 12 12

(12+12)×(5-4)

(5-12/4)×12

(12/12+5)×4

4 5 12 13

13+12+4-5

(13-5)×12/4

(13+5-12)×4

4 5 13 13

(13/13+5)×4

4 6 6 6

6×4+6-6

(6+6)×(6-4)

4 6 6 7

(7-6)×6×4

(7-4)×6+6

(7-6/6)×4

4 6 6 8

8+6+6+4

8×6-6×4

8×6/(6-4)

6×6-8-4

(6+6)×8/4

(6-8/4)×6

(4-6/6)×8

4 6 6 9

(6-4)×9+6

9×4-6-6

(9-4)×6-6

4 6 6 10

(10+6)×6/4

4 6 6 12

(6-4)×6+12

12×6/4+6

4 6 7 7

7+7+6+4

6×4+7-7

4 6 7 8

(6+4-7)×8

(8-7)×6×4

4 6 7 9

(9+7)×6/4

4 6 7 10

(6-4)×7+10

7×4+6-10

(7-4)×10-6

(10-6)×7-4

4 6 7 12

12×6/(7-4)

(7-12/4)×6

4 6 8 8

    6×4+8-8

    (6-4)×8+8

    (8+8)×6/4

    (8+4)×(8-6)

4 6 8 9

    (9-8)×6×4

    9×8/4+6

    (4-8/6)×9

4 6 8 10

    (8-6)×10+4

    (10-6)×4+8

4 6 8 12

    (6+4-8)×12

    8×6/4+12

    (8+4)×12/6

    (12-6)×(8-4)

    12×4/(8-6)

    (6-12/4)×8

    (8-12/6)×4

4 6 8 13

    (13-6-4)×8

4 6 9 9

    6×4+9-9

4 6 9 10

    (10-9)×6×4

    10×6/4+9

    10×6-9×4

    (10×9+6)/4

4 6 9 12

    (12+4)×9/6

    (12-4)×(9-6)

    (9-6)×4+12

4 6 9 13

    13+9+6-4

4 6 10 10

6×4+10-10

10×4-10-6

(10-4)×(10-6)

4 6 10 11

(11-10)×6×4

4 6 10 12

12+10+6-4

(10-4)×6-12

12×10/4-6

12×10/6+4

4 6 11 11

6×4+11-11

11+11+6-4

4 6 11 12

(12-11)×6×4

4 6 12 12

(12-6-4)×12

6×4+12-12

12×6-12×4

(4-12/6)×12

4 6 12 13

(13-12)×6×4

4 6 13 13

6×4+13-13

13+13+4-6

4 7 7 7

(7-7/7)×4

4 7 7 8

(7+7-8)×4

(4-7/7)×8

4 7 7 11

7×4+7-11

(11-7)×7-4

4 7 8 8

(7+4-8)×8

8×7-8×4

(7-8/8)×4

4 7 8 9

9×8/(7-4)

(8+4)×(9-7)

(8+7-9)×4

4 7 8 10

8×7/4+10

4 7 8 11

(11-7)×4+8

4 7 8 12

7×4+8-12

(12-8)×7-4

4 7 8 13

13+8+7-4

(13-7)×(8-4)

4 7 9 9

(7-9/9)×4

4 7 9 10

10×4-9-7

(9+7-10)×4

(9-7)×10+4

4 7 9 11

(7-4)×11-9

(9-4)×7-11

4 7 9 12

(7+4-9)×12

12+9+7-4

12×4/(9-7)

4 7 9 13

7×4+9-13

(13-9)×7-4

4 7 10 10

(7-10/10)×4

4 7 10 11

11+10+7-4

(10-4)×(11-7)

(10+7-11)×4

4 7 10 12

(10+4)×12/7

(12-4)×(10-7)

(10-7)×4+12

4 7 11 11

(7-11/11)×4

4 7 11 12

(11+7-12)×4

4 7 11 13

11×4-13-7

4 7 12 12

(7-4)×12-12

(12×7+12)/4

(7-12/12)×4

4 7 12 13

(13-7-4)×12

(12+7-13)×4

4 7 13 13

(7-13/13)×4

4 8 8 8

(8-4)×8-8

8×8/4+8

(4-8/8)×8

4 8 8 9

(8+4-9)×8

4 8 8 10

(8+4)×(10-8)

10×4-8-8

8×8-10×4

(8+8-10)×4

4 8 8 11

(11×8+8)/4

4 8 8 12

12+8+8-4

12×8/(8-4)

(12-8)×4+8

4 8 8 13

(13×8-8)/4

4 8 9 9

(4-9/9)×8

4 8 9 10

(9+4-10)×8

4 8 9 11

(8+4)×(11-9)

11+9+8-4

(9+8-11)×4

4 8 9 12

9×8×4/12

(8-4)×9-12

9×8-12×4

4 8 9 13

8/4+13+9

(13-9)×4+8

4 8 10 10

10+10+8-4

(10-8)×10+4

(4-10/10)×8

4 8 10 11

(10+4-11)×8

4 8 10 12

(8+4)×(12-10)

(8+4-10)×12

8/4+12+10

(10-4)×(12-8)

12×4/(10-8)

(10+8-12)×4

4 8 11 11

8/4+11+11

(4-11/11)×8

4 8 11 12

(11+4-12)×8

11×4-12-8

(12-4)×(11-8)

(11-8)×4+12

4 8 11 13

(8+4)×(13-11)

(11+8-13)×4

4 8 12 12

(12+4)×12/8

(4-12/12)×8

4 8 12 13

(12+4-13)×8

12/4+13+8

4 8 13 13

13+13-8/4

(4-13/13)×8

4 9 9 10

10+9+9-4

4 9 9 12

(9+9-12)×4

(4-12/9)×9

4 9 10 11

(11-9)×10+4

4 9 10 12

12×10/(9-4)

4 9 10 13

(10-4)×(13-9)

(10+9-13)×4

4 9 11 11

11×4-11-9

4 9 11 12

(9+4-11)×12

12×11/4-9

12×4/(11-9)

4 9 12 12

(12-4)×(12-9)

12/4+12+9

(12×9-12)/4

(12-9)×4+12

4 10 10 11

11×4-10-10

4 10 10 12

(12-10)×10+4

4 10 11 12

12/4+11+10

4 10 11 13

(13-11)×10+4

4 10 12 12

(10+4-12)×12

12×12/(10-4)

12×4/(12-10)

4 10 12 13

(12-4)×(13-10)

(13-10)×4+12

4 11 12 13

(11+4-13)×12

12×4-13-11

12×4/(13-11)

4 12 12 12

12×4-12-12

12×12/4-12

5 5 5 5

5×5-5/5

5 5 5 6

5×5+5-6

(5-5/5)×6

5 5 5 9

9+5+5+5

5 5 5 12

(5+5)×12/5

5 5 6 6

(5+5-6)×6

5×5-6/6

(6-6/5)×5

5 5 6 7

5×5+6-7

7×5-6-5

5 5 6 8

8+6+5+5

5 5 6 11

6×5+5-11

(11-5)×5-6

5 5 7 7

7+7+5+5

7×7-5×5

5×5-7/7

(7+5)×(7-5)

5 5 7 8

(5+5-7)×8

5×5+7-8

5 5 7 10

(7+5)×10/5

5 5 7 11

(7-11/5)×5

5 5 8 8

5×5-8/8

5 5 8 9

5×5+8-9

(8-5)×5+9

5 5 8 10

(10+5)×8/5

(5-10/5)×8

5 5 8 11

8×5-11-5

5 5 8 12

(5+5-8)×12

5 5 8 13

(13-5-5)×8

(13-5)×(8-5)

5 5 9 9

  5×5-9/9

5 5 9 10

  5×5+9-10

5 5 9 11

  (11-5)×(9-5)

5 5 10 10

  5×5-10/10

5 5 10 11

  5×5+10-11

5 5 10 13

  5/5+13+10

  (5-13/5)×10

5 5 11 11

  5×5-11/11

5 5 11 12

  5×5+11-12

  5/5+12+11

  (12-5)×5-11

5 5 11 13

  13+11+5-5

5 5 12 12

  (12-5-5)×12

  5×5-12/12

  12+12+5-5

5 5 12 13

  5×5+12-13

  13+12-5/5

5 5 13 13

  5×5-13/13

5 6 6 6

  (5-6/6)×6

5 6 6 7

  7+6+6+5

  (6+5-7)×6

  6×6-7-5

(6+6)×(7-5)

5 6 6 8

(8-5)×6+6

5 6 6 9

9×6-6×5

5 6 6 10

(10-5)×6-6

(6+6)×10/5

(6-10/5)×6

5 6 6 12

6×5+6-12

12×5-6×6

(12-6)×5-6

5 6 7 7

(5-7/7)×6

5 6 7 8

(7+5)×(8-6)

(7+5-8)×6

8×6/(7-5)

5 6 7 9

(7-5)×9+6

5 6 7 12

(7+5)×12/6

(7-5)×6+12

5 6 7 13

6×5+7-13

(13+7)×6/5

7×6-13-5

(13-7)×5-6

5 6 8 8

(6+5-8)×8

(5-8/8)×6

5 6 8 9

(8+5-9)×6

(9+6)×8/5

5 6 8 10

8×6×5/10

8×5-10-6

(8-5)×10-6

5 6 8 12

(12+8)×6/5

12×6/(8-5)

(5-12/6)×8

5 6 8 13

(13+5)×8/6

5 6 9 9

(9-6)×5+9

(5-9/9)×6

5 6 9 10

(9+5-10)×6

10×9/5+6

5 6 9 11

(11+9)×6/5

(11+5)×9/6

5 6 9 12

(6+5-9)×12

(12-6)×(9-5)

5 6 9 13

(13-5)×(9-6)

5 6 10 10

(10+10)×6/5

(5-10/10)×6

5 6 10 11

(10+5-11)×6

(11-5)×(10-6)

5 6 10 12

10×6/5+12

(6-12/10)×5

5 6 10 13

13+10+6-5

5 6 11 11

(5-11/11)×6

5 6 11 12

    12+11+6-5

    (11+5-12)×6

    (11-5)×6-12

5 6 11 13

    (13+11)×(6-5)

    (13-6)×5-11

5 6 12 12

    (12+12)×(6-5)

    (5-12/12)×6

5 6 12 13

    (13-6-5)×12

    13+12+5-6

    (12+5-13)×6

5 6 13 13

    (5-13/13)×6

5 7 7 9

    (7+5)×(9-7)

5 7 7 10

    (7-5)×7+10

5 7 7 11

    (5-11/7)×7

5 7 8 8

    (7-5)×8+8

    (8+7)×8/5

5 7 8 9

    (7+5-9)×8

    8×5-9-7

5 7 8 10

    (7+5)×(10-8)

5 7 9 10

    (10-7)×5+9

5 7 9 11

    (7+5)×(11-9)

5 7 9 12

    (9+5)×12/7

5 7 9 13

    13+9+7-5

    (13-7)×(9-5)

5 7 10 10

    10×7/5+10

5 7 10 11

    (10-5)×7-11

5 7 10 12

    (7+5)×(12-10)

    (7+5-10)×12

    12+10+7-5

5 7 10 13

    (13-5)×(10-7)

5 7 11 11

    11+11+7-5

    (11-5)×(11-7)

5 7 11 13

    (7+5)×(13-11)

5 7 12 12

    12×7-12×5

5 7 13 13

    13+13+5-7

5 8 8 8

    8×5-8-8

    8×8-8×5

5 8 8 9

    9×8/(8-5)

    (9-5)×8-8

5 8 8 10

    (8+5-10)×8

    10×8/5+8

5 8 8 13

    13+8+8-5

5 8 9 11

    (8-5)×11-9

    (9+5-11)×8

(11-8)×5+9

5 8 9 12

12+9+8-5

12×8/(9-5)

5 8 9 13

9×5-13-8

5 8 10 11

11+10+8-5

5 8 10 12

(10+5-12)×8

5 8 11 12

(8+5-11)×12

(11+5)×12/8

(11-5)×(12-8)

5 8 11 13

(11+5-13)×8

(13-5)×(11-8)

5 8 12 12

(8-5)×12-12

5 9 9 11

11+9+9-5

5 9 9 12

9×5-12-9

(9-5)×9-12

(12-9)×5+9

5 9 10 10

10+10+9-5

5 9 10 11

9×5-11-10

5 9 10 13

10/5+13+9

(13-10)×5+9

5 9 11 13

(11-5)×(13-9)

5 9 12 12

(9+5-12)×12

(12×9+12)/5

5 9 12 13

(13+5)×12/9

(13-5)×(12-9)

5 10 10 11

(11×10+10)/5

5 10 10 12

12×10/(10-5)

10/5+12+10

5 10 10 13

(13×10-10)/5

5 10 11 11

10/5+11+11

5 10 12 13

(10+5-13)×12

5 10 13 13

10×5-13-13

13+13-10/5

(13-5)×(13-10)

5 11 12 12

12×12/(11-5)

(12×11-12)/5

6 6 6 6

6+6+6+6

6×6-6-6

6 6 6 8

(6+6)×(8-6)

(6+6-8)×6

6 6 6 9

6×6×6/9

(9-6)×6+6

6 6 6 10

10×6-6×6

6 6 6 11

(11-6)×6-6

6 6 6 12

(6+6)×12/6

(6-12/6)×6

6 6 7 9

(6+6)×(9-7)

(7+6-9)×6

6 6 7 10

(10-7)×6+6

6 6 7 11

11×6-7×6

6 6 7 12

7×6-12-6

(12-7)×6-6

6 6 8 8

8×6/(8-6)

6 6 8 9

(6+6-9)×8

(8-6)×9+6

6 6 8 10

(6+6)×(10-8)

(8+6-10)×6

6 6 8 11

(11-8)×6+6

6 6 8 12

8×6×6/12

12×6-8×6

(8-6)×6+12

(12+6)×8/6

6 6 8 13

(13-8)×6-6

6 6 9 10

(9-6)×10-6

(10+6)×9/6

6 6 9 11

(6+6)×(11-9)

(9+6-11)×6

6 6 9 12

12×6/(9-6)

12×9/6+6

(12-9)×6+6

6 6 9 13

13×6-9×6

6 6 10 12

(6+6)×(12-10)

(6+6-10)×12

(10+6-12)×6

(12-6)×(10-6)

6 6 10 13

6/6+13+10

(13-10)×6+6

6 6 11 12

6/6+12+11

6 6 11 13

(6+6)×(13-11)

13+11+6-6

(11+6-13)×6

6 6 12 12

12+12+6-6

(12-6)×6-12

6 6 12 13

13+12-6/6

6 7 7 10

(7+7-10)×6

6 7 7 11

7×6-11-7

6 7 8 9

8×6/(9-7)

6 7 8 10

(7+6-10)×8

7×6-10-8

(8-6)×7+10

6 7 8 11

(11+7)×8/6

(8+7-11)×6

6 7 8 12

(8+6)×12/7

6 7 9 9

7×6-9-9

(9+7)×9/6

(9-7)×9+6

6 7 9 12

(9+7-12)×6

(9-7)×6+12

6 7 10 10

(10-7)×10-6

6 7 10 12

12×7/6+10

12×7-10×6

12×6/(10-7)

6 7 10 13

13+10+7-6

(13-7)×(10-6)

(10+7-13)×6

6 7 11 11

(11-6)×7-11

6 7 11 12

(7+6-11)×12

12+11+7-6

(12-6)×(11-7)

6 7 11 13

(13+11)×(7-6)

6 7 12 12

(12+12)×(7-6)

6 7 12 13

13+12+6-7

(13-7)×6-12

6 8 8 8

(8-6)×8+8

6 8 8 9

9×8-8×6

(8+8)×9/6

6 8 8 10

8×6/(10-8)

(10+8)×8/6

(10-6)×8-8

6 8 8 11

(8+6-11)×8

6 8 8 12

12×8/6+8

(8+8-12)×6

6 8 9 9

(9+9)×8/6

9×8/(9-6)

6 8 9 10

(10-8)×9+6

6 8 9 11

8×6/(11-9)

6 8 9 12

9×8/6+12

(9+6-12)×8

6 8 9 13

13+9+8-6

(9+8-13)×6

6 8 10 11

(11-8)×10-6

6 8 10 12

8×6/(12-10)

12+10+8-6

(10+6)×12/8

12×8/(10-6)

(10-8)×6+12

6 8 10 13

(10+6-13)×8

6 8 11 11

11+11+8-6

6 8 11 12

12×6/(11-8)

6 8 11 13

8×6-13-11

8×6/(13-11)

6 8 12 12

(8+6-12)×12

8×6-12-12

12×8-12×6

(12-6)×(12-8)

12×12/8+6

6 8 13 13

13+13+6-8

6 9 9 10

10×9/6+9

6 9 9 11

(9-6)×11-9

(11-9)×9+6

6 9 9 12

12+9+9-6

6 9 10 11

11+10+9-6

10×9-11×6

6 9 10 12

(10-6)×9-12

(12-9)×10-6

(12-10)×9+6

6 9 11 12

(11-9)×6+12

6 9 11 13

(13-11)×9+6

6 9 12 12

(9-6)×12-12

(12+6)×12/9

12×6/(12-9)

6 9 12 13

(9+6-13)×12

(12-6)×(13-9)

12/6+13+9

6 10 10 10

10+10+10-6

6 10 10 13

(13-10)×10-6

6 10 11 12

12×10/(11-6)

6 10 12 12

12/6+12+10

(12-10)×6+12

6 10 12 13

12×6/(13-10)

6 11 11 12

12/6+11+11

6 11 12 12

(12×11+12)/6

6 11 12 13

(13-11)×6+12

6 12 12 12

12×12/(12-6)

6 12 12 13

(13×12-12)/6

6 12 13 13

13+13-12/6

7 7 7 12

(7+7)×12/7

7 7 8 11

(7+7-11)×8

7 7 9 10

(9-7)×7+10

7 7 10 13

7/7+13+10

7 7 11 12

7/7+12+11

(12-7)×7-11

7 7 11 13

13+11+7-7

(13-7)×(11-7)

7 7 12 12

(7+7-12)×12

12+12+7-7

7 7 12 13

7×7-13-12

13+12-7/7

7 8 8 9

(9-7)×8+8

7 8 8 10

10×8-8×7

7 8 8 11

(11-7)×8-8

7 8 8 12

(8+7-12)×8

7 8 8 13

(13+8)×8/7

7 8 9 10

9×8/(10-7)

7 8 9 12

(12+9)×8/7

(9+7)×12/8

7 8 9 13

(9+7-13)×8

7 8 10 10

(10-8)×7+10

7 8 10 11

(11+10)×8/7

7 8 10 13

13+10+8-7

7 8 11 12

12+11+8-7

12×8/(11-7)

7 8 11 13

    (13+11)×(8-7)

    (13-8)×7-11

7 8 12 12

    (12+12)×(8-7)

7 8 12 13

    (8+7-13)×12

    13+12+7-8

    (13-7)×(12-8)

7 9 9 13

    13+9+9-7

7 9 10 11

    (10-7)×11-9

    (11-9)×7+10

7 9 10 12

    12+10+9-7

7 9 11 11

    11+11+9-7

7 9 11 12

    (11+7)×12/9

    (11-7)×9-12

7 9 12 12

    12×9-12×7

7 9 13 13

    13+13+7-9

    (13-7)×(13-9)

7 10 10 11

    11+10+10-7

7 10 10 12

    (12-10)×7+10

7 10 11 13

    (13-11)×7+10

7 10 12 12

    (10-7)×12-12

    12×10/(12-7)

7 10 12 13

117

(13+7)×12/10

7 12 12 13

12×12/(13-7)

(13×12+12)/7

8 8 8 10

(10-8)×8+8

8 8 8 11

11×8-8×8

8 8 8 12

(8+8)×12/8

(12-8)×8-8

8 8 8 13

(8+8-13)×8

8 8 9 11

9×8/(11-8)

(11-9)×8+8

8 8 9 12

12×8-9×8

8 8 9 13

(13-9)×8-8

8 8 10 12

(12-10)×8+8

8 8 10 13

8/8+13+10

13×8-10×8

8 8 11 12

8/8+12+11

8 8 11 13

13+11+8-8

(13-11)×8+8

8 8 12 12

12+12+8-8

12×8/(12-8)

8 8 12 13

13+12-8/8

8 9 9 12

9×8/(12-9)

8 9 10 12

(10+8)×12/9

12×10/8+9

8 9 10 13

9×8/(13-10)

13+10+9-8

8 9 11 11

(11-8)×11-9

8 9 11 12

12+11+9-8

8 9 11 13

(13+11)×(9-8)

8 9 12 12

(12+12)×(9-8)

(12-8)×9-12

12×12/9+8

8 9 12 13

13+12+8-9

12×8/(13-9)

8 10 10 12

12+10+10-8

8 10 11 11

11+11+10-8

8 10 12 12

(12+8)×12/10

12×10-12×8

8 10 12 13

12×10/(13-8)

8 10 13 13

13+13+8-10

8 11 12 12

(11-8)×12-12

9 9 9 12

(9+9)×12/9

9 9 10 13

119

9/9+13+10

9 9 11 12

9/9+12+11

(12-9)×11-9

9 9 11 13

13+11+9-9

9 9 12 12

12+12+9-9

9 9 12 13

13+12-9/9

(13-9)×9-12

9 10 10 13

13+10+10-9

9 10 11 12

12+11+10-9

(11+9)×12/10

9 10 11 13

(13+11)×(10-9)

(13-10)×11-9

9 10 12 12

(12+12)×(10-9)

9 10 12 13

13+12+9-10

9 11 11 11

11+11+11-9

9 11 12 12

12×11-12×9

9 11 12 13

(13+9)×12/11

9 11 13 13

13+13+9-11

9 12 12 12

(12-9)×12-12

10 10 10 12

(10+10)×12/10

10 10 10 13

10/10+13+10

10 10 11 12

10/10+12+11

10 10 11 13

13+11+10-10

10 10 12 12

12+12+10-10

10 10 12 13

13+12-10/10

10 11 11 12

12+11+11-10

10 11 11 13

(13+11)×(11-10)

11/11+13+10

10 11 12 12

(12+12)×(11-10)

(12+10)×12/11

10 11 12 13

13+12+10-11

10 12 12 12

12×12-12×10

10 12 12 13

12/12+13+10

(13-10)×12-12

10 12 13 13

13+13+10-12

10 13 13 13

13/13+13+10

11 11 11 12

(11+11)×12/11

11/11+12+11

11 11 11 13

13+11+11-11

11 11 12 12

12+12+11-11

11 11 12 13

13+12-11/11

(13+11)×(12-11)

11 12 12 12

12/12+12+11

(12+12)×(12-11)

11 12 12 13

13+12+11-12

13×12-12×11

11 12 13 13

13/13+12+11

(13+11)×(13-12)

11 13 13 13

13+13+11-13

12 12 12 12

12+12+12-12

12 12 12 13

(12+12)×(13-12)

13+12-12/12

12 12 13 13

13+12+12-13

12 13 13 13

13+12-13/13

(13+13)×12/13